READING POWER

Earth Rocks!

Igneous Rocks

Holly Cefrey

The Rosen Publishing Group's
PowerKids Press™
New York

Published in 2003 by The Rosen Publishing Group, Inc.
29 East 21st Street, New York, NY 10010

First Edition

Book Design: Mindy Liu

Photo Credits: Cover, pp. 6–7 © Digital Vision; pp. 4–5 © Paul A. Souders/Corbis; p. 8 © Kevin Fleming/Corbis; pp. 9, 12 (inset) © Breck P. Kent/Animals Animals; p. 10 © Bill Ross/Corbis; p. 11 © James A. Sugar/Corbis; pp. 12–13 © Michael S. Yamashita/Corbis; pp. 14, 17 (top) © Yann Arthus-Bertrand/Corbis; p. 15 © Reuters NewMedia Inc./Corbis; p. 15 (inset) © MapArt; p. 17 (middle) © James L. Amos/Corbis; p. 17 (bottom) © Michelle Garrett/Corbis; pp. 18–19 © Roger Ressmeyer/Corbis; p. 18 (inset) © Roger W. Madden/National Geographic Image Collection; pp. 20–21 © James Watt/Animals Animals

Library of Congress Cataloging-in-Publication Data

Cefrey, Holly.
Igneous rocks / Holly Cefrey.
    p. cm. — (Earth rocks!)
Summary: Describes some of the properties and uses of igneous rocks.
Includes bibliographical references and index.
ISBN 0-8239-6464-7 (library binding)
1. Rocks, Igneous—Juvenile literature. [1. Rocks, Igneous.] I.
Title.
QE461 .C39 2003
552′.1—dc21

                                        2002000109

Contents

Earth Rocks

Igneous rocks are found all over Earth. There are more than 600 different kinds of igneous rocks. They are different colors, shapes, and sizes.

Granite can be found in several different colors, such as pink and gray.

The Fact Box

Some igneous rocks are 3.6 billion years old. The word *igneous* comes from the Latin word for "fire."

How Igneous Rocks Form

Igneous rocks are made above and below Earth's surface. Earth is made of layers of rock. The top layer is called the crust. Inside the crust is hot, melted rock, called magma. When magma cools, it hardens into igneous rocks.

Earth's Layers

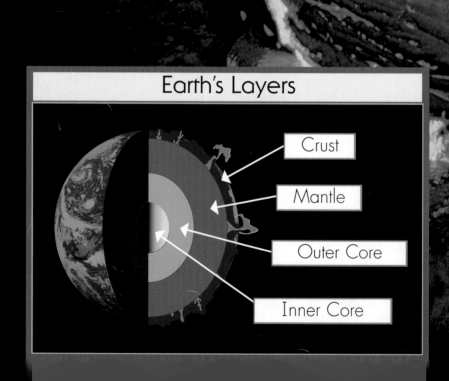

Crust

Mantle

Outer Core

Inner Core

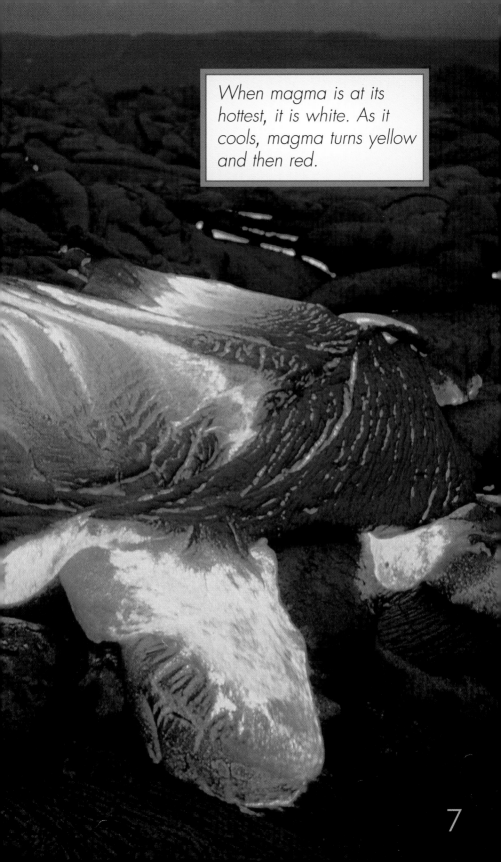

When magma is at its hottest, it is white. As it cools, magma turns yellow and then red.

Magma can take millions of years to cool and harden inside Earth. During this time, crystals form. You can see crystals when you look at igneous rocks, such as granite.

People must dig deep below Earth's surface to get granite. Granite is used for building many things.

Feldspar

Mica

Quartz

We can see the crystals of mica, feldspar, and quartz in this piece of granite.

Most igneous rocks that form above the surface of Earth are made by volcanoes. When a volcano erupts, hot lava comes out. The lava cools quickly and hardens into igneous rock.

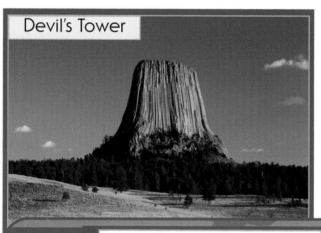

Devil's Tower

The Fact Box

Devil's Tower was formed when magma went up into a volcano but didn't come out of the top. The magma cooled and hardened inside the volcano. Over time, weather wore away the volcano's surface. A large column of igneous rock was left.

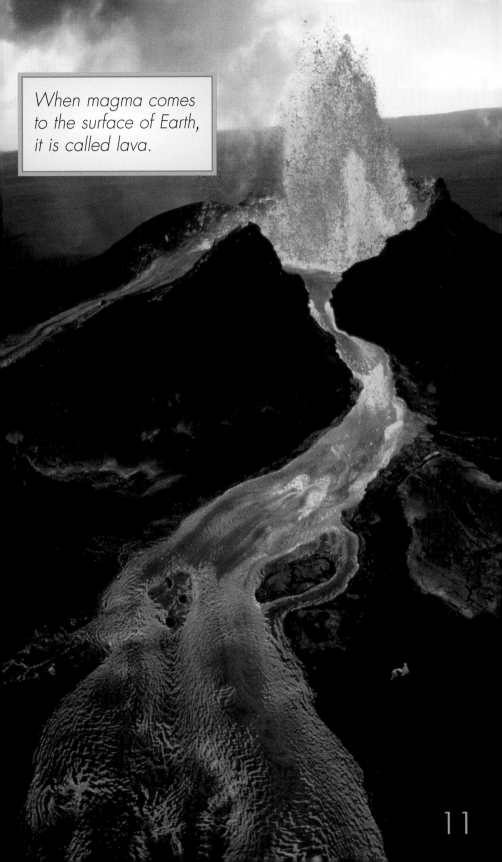

When magma comes to the surface of Earth, it is called lava.

11

Lava can make many different kinds of igneous rock. When lava has lots of air in it and is bubbly, pumice is formed. Pumice is filled with air holes. It is so light that it floats on water!

This is a close-up look at the holes in a piece of pumice.

Basalt is an igneous rock that forms when lava cools quickly. Most of Earth's crust is basalt. These many-sided columns of basalt are in Northern Ireland. They were formed about 30 million years ago.

Underwater Volcanoes

Sometimes, volcanoes erupt underwater. Lava from underwater volcanoes has made many islands on our planet. All of the Hawaiian Islands were made by volcanoes.

Surtsey

The Fact Box

In 1963, erupting underwater volcanoes formed a new island near Iceland called Surtsey. Many plants and birds now live there.

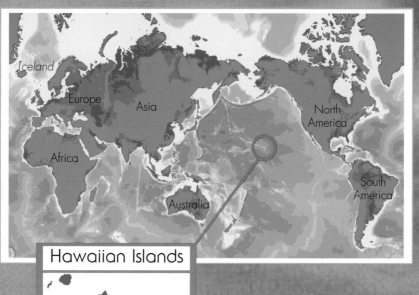

Iceland

Europe

Asia

Africa

Australia

North
America

South
America

Hawaiian Islands

An underwater volcano
makes a lot of steam
when it erupts.

Using Igneous Rocks

People have used igneous rocks for
thousands of years. Long ago, people
used obsidian to make tools. Today,
people use granite to make buildings.
Granite is not hurt by changes in the
weather. People use pumice to decorate
around trees and bushes. Pumice is
also used in some soaps and is used
for cleaning.

Empire State Building

Parts of one of the world's tallest buildings are made of granite.

Obsidian Arrowhead

Tools made from obsidian were very sharp.

Some people rub a pumice stone on the bottoms of their feet to make their skin smooth.

Pumice

Studying Igneous Rocks

A geologist is a person who studies rocks. Some geologists study magma and lava. Lava is very hot. It can reach 2,500 degrees Fahrenheit (1,371 degrees Celsius). It burns through most things. Geologists have to use special tools to collect lava.

Geologists must often get very close to erupting volcanoes to study them.

Geologists must be very careful not to get burned by hot lava.

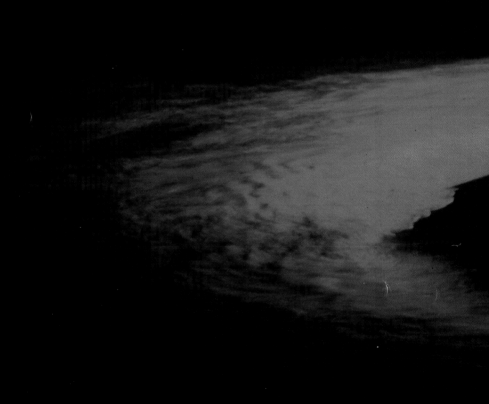

Igneous rocks are Earth's building blocks. They have formed some of our tallest mountains, biggest islands, and most powerful volcanoes. Today, igneous rocks are still forming and shaping our planet.

Lava forms new land as it cools and becomes igneous rock.

Glossary

basalt (buh-**sawlt**) the most common igneous rock made from lava

crust (**kruhst**) the solid, outer part of a planet

crystals (**krihs**-tlz) hard, usually clear, matter that has angles and flat surfaces

erupt (ih-**ruhpt**) when a volcano throws out rocks, hot ashes, and lava with great force

geologist (jee-**ahl**-uh-jihst) a person who studies Earth's rocks

igneous rock (**ihg**-nee-uhs **rahk**) rock that is made from magma or lava

lava (**lah**-vuh) hot, melted rock, or magma, that reaches Earth's surface

magma (**mag**-muh) melted, flowing rock that is found below Earth's surface

obsidian (ahb-**sihd**-ee-uhn) a hard, dark, glassy rock that forms from lava

pumice (**puhm**-ihs) a light rock formed from lava

Resources

Books

Eyewitness: Rocks and Minerals
by R. F. Symes
Dorling Kindersley Publishing (2000)

Rocks and Minerals
by Neil Morris
Crabtree Publishing (1999)

Web Sites

Due to the changing nature of Internet links, PowerKids
Press has developed an on-line list of Web sites related
to the subjects of this book. This site is updated regularly.
Please use this link to access the list:

http://www.powerkidslinks.com/ear/ign/

Index

Word Count: 392

Note to Librarians, Teachers, and Parents

If reading is a challenge, Reading Power is a solution! Reading Power is perfect for readers who want high-interest subject matter at an accessible reading level. These fact-filled, photo-illustrated books are designed for readers who want straightforward vocabulary, engaging topics, and a manageable reading experience. With clear picture/text correspondence, leveled Reading Power books put the reader in charge. Now readers have the power to get the information they want and the skills they need in a user-friendly format.